D0622854

JUN - 2006

Volcanologist

Animal Therapist

Astrobiologist

Computer Game Developer

Pyrotechnician

SETI Scientist

Virus Hunter

Volcanologist

WEIRD
CAREERS
in SCIENCE

Volcanologist

Mary Firestone

CHELSEA HOUSE
PUBLISHERS
A Haights Cross Communications Company ®
Philadelphia

CHELSEA HOUSE PUBLISHERS

VP, NEW PRODUCT DEVELOPMENT Sally Cheney
DIRECTOR OF PRODUCTION Kim Shinners
CREATIVE MANAGER Takeshi Takahashi
MANUFACTURING MANAGER Diann Grasse
SERIES DESIGNER Takeshi Takahashi
COVER DESIGNER Takeshi Takahashi

STAFF FOR **VOLCANOLOGIST**

PROJECT MANAGEMENT Ladybug Editorial and Design
DEVELOPMENT EDITOR Tara Koellhoffer
LAYOUT Gary Koellhoffer

A Haights Cross Communications ✦ Company ®

www.chelseahouse.com

First Printing

9 8 7 6 5 4 3 2 1

Library of Congress Cataloging-in-Publication Data

Firestone, Mary.
  Volcanologists / Mary Firestone.
    p. cm. — (Weird careers in science)
  Includes bibliographical references and index.
  ISBN 0-7910-8702-6
  1. Geology—Vocational guidance—Juvenile literature. 2. Volcanolo-
gists—Juvenile literature. 3. Volcanoes—Juvenile literature. I. Title. II.
Series.
  GE34.F57 2005
  551'.023—dc22

                                    2005012077

TABLE OF CONTENTS

Chapter 1

Volcanoes and the People Who Study Them

IN 1993, volcanologist Dr. Stanley Williams was standing inside Galeras, an active volcano located in Colombia, South America (Figure 1.1). Steam was rising around him, and gases were escaping from nearby **fumeroles**. He had tested the gases and heat from the volcano earlier, and had determined that it was safe to go inside the volcano. A team of other volcanologists and a few tourists were with him, taking photographs and exploring the area. Suddenly, the volcano erupted. Flying rocks, **lava**, and fire killed six people instantly. Dr. Williams ran from the explosion, his backpack in flames. Two of his fellow volcanologists were able to reach him through the smoke,

Figure 1.1 In 1993, Mount Galeras in Colombia erupted, almost killing volcanologist Stanley Williams.

and they pulled him to safety. Severely injured, Williams survived to tell his story. But this incident was a strong reminder of just how dangerous his job could be.

Volcanology is a science devoted to the most powerful explosive force on Earth—volcanic eruptions. Volcanologists love the challenge and the thrill of studying dangerous, active volcanoes. Their work brings them close to the awesome power of nature, and also helps them save thousands of lives.

WHAT CAUSES VOLCANOES?

A volcano is an opening in the Earth's crust, where volcanic eruptions have occurred. Volcanic eruptions occur when **molten** rock called **magma** (magma is called "lava" until it reaches the opening of the volcano) rises up through these openings in the Earth's crust. These openings exist mostly along the spaces between massive rock plates, which scientists call **tectonic plates**. These plates fall into two categories: oceanic (forming the ocean floor) and continental (forming the land). Together, they form the **lithosphere**.

These massive plates are thousands of miles across, but there are some smaller plates, too. Volcanologists have determined that there are seven major plates and several smaller ones (Figure 1.2). The large plates are as big as a continent, "but most commonly there are small plates about the size of a small country," says University of North Dakota volcanologist Shan de Silva.

The edges of continental and oceanic plates collide from time to time, because the Earth's crust is always moving and shifting, heating and cooling, and adjusting itself. The collision of two or more rock plates causes one plate to give

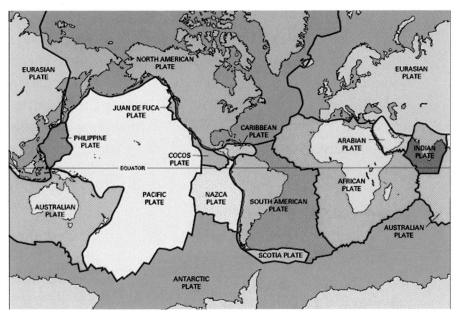

Figure 1.2 The surface of the Earth is covered with several large tectonic plates and many smaller ones, as seen on this map.

way, and sink below another, melting in the hot **mantle** layer as it dives down. These areas along the edges of continental plates and oceanic plates, where one plate gives way to another, are called "**subduction zones**." Many volcanoes exist along these zones. A famous subduction zone volcano is Mount St. Helens, in the Cascade Range. The Cascade Range is located along the west coast of North America, from British Columbia, down through Washington and Oregon, and northern California.

WHERE DO VOLCANOES ERUPT?

Volcanoes don't just pop up anywhere. Scientists have learned that most of the planet's active volcanoes are found over the edges, or ridges, of the oceanic and continental plates.

Most of the world's volcanoes exist along the Pacific Rim, which is the edge of an oceanic plate. It forms a huge arc shape in the Pacific Ocean. This arc is called the Ring of Fire, and it is dotted with volcanoes. The Ring of Fire runs along the ocean plate that borders the west coasts of South America, Mexico, the United States (including Alaska) and Canada, the Aleutian Islands, eastern Russia, Japan, and New Zealand (Figure 1.3). Scientists say there are 600 active volcanoes in this area alone, and many more **dormant** (inactive) ones as well.

HOW VOLCANOES FORM

As scientists studied volcanoes, they noticed that they occurred in four different ways on the Earth's crust.

Ocean Ridge Volcanoes

Ocean ridge volcanoes occur along the edges of tectonic plates on the floor of the Atlantic Ocean. **Fissures** mark the

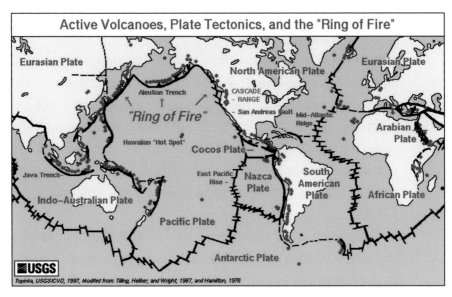

Figure 1.3 The Ring of Fire around the rim of the Pacific Ocean, seen here at left center, contains some of the most active volcanoes in the world.

areas where different plates are moving away from each other. Magma rises through the fissures and erupts onto the ocean floor, and onto edges of each side of the fissure, forming mountains and **black smokers**. Black smokers are a type of vent found on the ocean floor. They are hundreds of meters wide, and are formed when steaming-hot water beneath the Earth's crust comes through the ocean floor.

Island Arc Volcanoes

Island arc volcanoes occur over oceanic plate subduction zones, where one oceanic plate collides with another. Island arc volcanoes are created more quickly than others, since they form in the ocean, where the magma doesn't have to force its way through **granite**. Island arc volcanoes create volcanic islands, such as the Aleutian Islands and Japan. These volcanoes appear in arc-shaped rows.

Continental Rift and Continental Margin Volcanoes

Have you ever wondered how coastal mountain ranges and new oceans are formed? When volcanic rock, which makes up the ocean floor, moves toward continental rock, it pushes up against it. Gradually, over millions of years, the underlying volcanic rock is pushed and crumpled upward, eventually rising to form a mountain chain in subduction zones. This is where continental margin volcanoes are formed.

Continental rifts are depressions in the land, caused by the action of rock below. In some cases, if a continental rift is deep enough, eventually a new ocean will form. In the Great Rift Valley of eastern Africa, a new ocean is currently being formed due to continental rifts. Scientists estimate that several million years from now, the Great Rift Valley will be separated from Africa by an ocean. This is the process that formed South America, which was separated from Africa 130 million years ago.

Hot Spot Volcanoes

The only kind of volcano that doesn't occur near the edge of a tectonic plate is a hot spot volcano. A "hot spot" is a super-hot area below the Earth's crust in the mantle. It regularly melts through the crust, pouring hot magma into the ocean. As the ocean plates shift and move over millions of years, they create rows of volcanic rock. Hot spot volcanoes created the Hawaiian and Galapagos Islands.

WHAT IS A VOLCANOLOGIST?

A volcanologist is a scientist who studies volcanoes. Volcanologists study all aspects of volcanoes: old volcanoes, new eruptions, the insides of volcanoes, and the flow of lava.

Volcanology is a fairly new science. When people first began to study volcanoes, they simply observed the surface of the Earth and the rocks that made up volcanoes. During the 17th and 18th centuries, the field grew into a complicated science that involved predicting and understanding the intimate relationships between plate tectonics and the composition of Earth elements such as water, gases, and different types of rock. As scientists learned more about the physical sciences, including **electromagnetism**, **gravity**,

Types of Volcanic Eruptions

One way that volcanologists classify volcanoes is by how often they erupt. There are four kinds:

1. **Active volcanoes** erupt regularly. They are generally quiet, but are occasionally violent. Stromboli is a famous active volcano in Italy. It has been erupting almost continually for 2,000 years. Some volcanologists estimate that it has been active for as long as 5,000 years. Most of the **cone** of Stromboli was developed 15,000 years ago.
2. **Intermittent volcanoes** erupt at regular periods, which makes predicting their eruptions a lot easier. Mount Asama in Japan is an intermittent volcano.
3. **Dormant volcanoes** are inactive, but not dead. They are "sleeping," but will likely erupt again one day. Lassen Peak in California and Paricutin in Mexico are considered dormant volcanoes.
4. **Extinct volcanoes** are volcanoes that have not been active at all since the beginning of recorded history. Crater Lake in Oregon is part of an extinct volcano.

and the solid structure of the Earth, volcanologists had more information to work with.

Better instruments were developed that helped volcanologists detect changes in the Earth and analyze what those changes meant. In the early days of volcano study, scientists used survey equipment, hammers, and compasses. Today, the tools of the trade include electronic sensing and analyzing equipment. Over time, volcanologists have become like scientist-detectives, who examine every tiny detail found in volcanic **deposits**. Computers also play a big role in volcanology. Volcanologists can create computerized models of eruptions, which help them understand why and how they happen.

In recent years, volcanologists have begun to focus more on chemical (gases and magma) and physical (electromagnetic fields, gravitational influences) processes that cause eruptions, and what type of deposits eruptions leave behind. This process has led to some new subdisciplines within volcanology.

A volcanologist might also be a geochemist, a geophysicist, computer analyst, physical volcanologist, a geodesist, seismologist, stratigrapher, mineralogist, or engineering geologist. Whatever the volcanologist's subdiscipline, he or she knows a lot about geology, and applies this knowledge to studying volcanoes.

Chapter 2

What Do Volcanologists Do?

WHEN A VOLCANOLOGIST comes home after a hard day's work, he or she might have some wild stories to tell, to say the very least. A volcanologist's job is a study in extremes: life-threatening excursions up the slope of an active, gassy volcano to collect samples from orange-red lava flows; or sitting at the computer, analyzing equations and charting graphs. But the level of risk taken by volcanologists is up to them. No one forces them to take the risks they do.

Before and after the eruption of Mount St. Helens in 1980, volcanologists dotted its flanks and the surrounding landscape with their instruments, obsessed with its activity. They were taking advantage of every opportunity to learn and apply the technological advances of computer models and electronic sensing equipment. However, this work sometimes involved danger, while they collected material from the volcano itself.

On the day after it erupted, Mount St. Helens continued to have smaller eruptions, spewing gray and black clouds, or "columns," as scientists call them. The volcanologists weren't sure what the columns were made of, but they needed to find out. If the columns were dark because of pulverized old rock, it meant that the eruption of the volcano did not involve magma. However, if newer, fresh rock was in the clouds, this meant that magma was closer to the surface and that these eruptions were a sign of a bigger eruption that was still to come.

The only way for scientists to learn what was in those clouds of smoke was to get close to the erupting material, at the **vent**. Don Swanson, a United States Geological Survey (USGS) scientist, decided he was the man for the job.

Volcano Bombs

When a volcano erupts, magma, dust, gas, steam, and hot rocks fly up out of the crater. Magma pouring out of the crater and down the sides of the volcano is called lava. Some of the magma cools on its way to the crater and forms solid pieces, which volcanologists call "bombs." A volcanic bomb is a flaming lump of **viscous** material. Types of bombs are named according to their texture. There are breadcrust bombs, ribbon bombs, spindle bombs, spheroidal bombs, and "cow-dung" bombs. Some of these bombs are as big as trucks and cars. Volcanic **ash** is about the size of marbles. The smallest stuff that erupts is volcanic dust.

He went to a hardware story in downtown Vancouver, Washington, and bought a yardstick and a soup ladle.

Writer Dick Thompson, in his 2000 book *Volcano Cowboys*, described what happened next:

> He [Swanson] taped the ladle to the end of the yardstick, drove an hour north to a baseball field near the volcano where the [USGS helicopter] was parked, and told the pilot to fly up to the side of the erupting mountain. The two men hovered just below the summit while trying to get a sense of the rhythm of the eruptions, and when one blast stopped, the helicopter darted to the mountain top. In high winds, the former Vietnam pilot held the chopper's skids inches from the lip of the **crater**. Swanson pushed open his door, leaned out and began scooping fresh dark ash.

Don Swanson and his helicopter pilot were lucky to live through this experience, and to return successfully with the sample. After examining it, Swanson and other scientists determined that the sample did not contain fresh magma and rock. To everyone's relief, that meant that another magma-driven explosion was *not* going to be happening soon.

IT'S ABOUT THE SCIENCE

Whether volcanologists are in the field swinging from helicopters or staring at a computer screen, their day usually involves a certain amount of crunching numbers. After all, this is a science. They have to analyze and interpret data from all kinds of samples, **seismometers**, and sensors. Sometimes, they take the results and test new ways of analyzing numbers, which may teach them something new.

Volcanologists do a lot of research while they investigate the history and origin of volcanoes and volcano-related

Figure 2.1 **Volcanologists often use helicopters to get close to the mouth of a volcano to collect samples of lava and gases.**

earthquakes. They read scientific journals and search the Internet for information. Another part of their job is constructing and interpreting maps, graphs, and diagrams.

They use computers to check on volcanic surveillance data from satellites. Volcanologists also prepare environmental impact studies, and test new theories about possible volcanic activity, both on the computer and in the laboratory.

PREDICTING VOLCANIC ERUPTIONS

Volcanologists study volcanoes because they love their awesome power and beauty. Studying volcanoes also helps them predict eruptions with greater accuracy, which can save the lives of people who may live in the path of an erupting volcano.

To do this, volcanologists visit extinct volcanoes, and apply what they learn about them to active volcanoes. They visit dormant and active volcanoes, bringing portable observatories equipped with sensors and computers, to get a sense of when the volcanoes might blow up.

Predicting the eruption of a volcano requires knowing a volcano's unique history, its current seismic and ground **deformation levels**, and the concentration of chemicals and gas emissions at the site.

When volcanologists arrive at a site, they walk around and check for cracks in the ground, or signs of rock slides. They set up a range of instruments over a wide area: seismometers to measure Earth tremors and the movement of magma, **magnetometers** to assess changes in the magnetic fields beneath the volcano, and **spectrometers** for monitoring the extent and content of volcanic gases. They usually monitor the results from a remote location.

Volcanologists measure temperatures of lava and gas, and collect samples (Figure 2.1). They measure the height of **lava fountains** and **ash plumes**; they check the flow rate of ashes being ejected, and the flow rate of lava. They use this and other information to come up with a behavior model of a particular volcano, or type of eruption, to understand how other, similar, volcanoes might behave, or to predict larger eruptions of an already erupting volcano.

HISTORY

The history of a volcano is understood by reading about past eruptions and studying its geology, such as the type of deposits of rock and ash that were formed by past eruptions. Volcanologists then try to reconstruct the history of a

Figure 2.2 Volcanologists often set up camp near a volcano, so they can take samples and photographs to measure the volcano's activity.

volcano with this information. When they understand a volcano's past, they can make some predictions about what it might do in the future. Layers of rock reveal the age of volcanic material and how often the volcano has erupted, even over thousands of years.

Learning the history of the volcano could also involve taking photographs of the area around the volcano and making maps of the kinds of rocks found near and inside the volcano (Figure 2.2). Volcanologists might also take samples of rock and lava, testing them for chemical makeup and dating each sample.

GAS ANALYSIS

Volcanoes give off poisonous, smelly gases, such as **sulfur**, chlorine, and fluorine, before, during, and after an eruption. Gas levels in magma are studied closely by geochemists at eruption sites, because high quantities of magmatic gas are a sign of a coming eruption. Volcanologists go back to the same volcano many times to collect samples of gas coming out of the ground or from fumeroles. They also use gas detectors to check for slight increases in gas levels, and test water temperature and soil acidity.

STUDYING GROUND DEFORMATION

Like water filling up a balloon, magma rises and pushes against the Earth's crust, forcing it outward, causing it to change shape and eventually to blow up. When magma rises further, up toward the crater, it pushes large areas of soil and rock upward, like a swelling, creating a dome. When these swellings are very large, volcanologists know that an eruption is coming. These ground deformations are important ways to predict and prevent disasters.

Once they have studied the history and current activities, volcanologists can create hazard assessment maps. These maps spell out the expected areas where lava and ash will flow during and after an eruption, and how soon the next eruption will occur.

WAYS TO BE A VOLCANOLOGIST
Subdisciplines
Geochemists
Geochemists study the chemical properties of soil, lava, and rocks at the site of a volcano. They closely study **trace elements** in magma, vapors, and gases for noxious (poiso-

nous) fumes and sulfur content. They study the volatile (explosive) elements in magma. They spend hours taking samples of highly noxious volcanic gas inside fumeroles, which are extremely hot. They bring these samples to laboratories to be analyzed.

The geochemist's work is extremely important to predicting when volcanoes will erupt. Studies have shown that the chemical balance of magma changes just before an eruption. Geochemists measure the quantity of gas a volcano gives off into the atmosphere. To do this, they install spectrometers at the site, and on satellites and aircraft.

Geophysicists

Geophysicists closely study the **seismic activity** around a volcano. Seismic activity is caused by differences in the way rocks move above and around the magma. These movements can be recorded, measured, and analyzed, to help scientists figure out how far down the magma goes into the Earth's crust, and how far it has moved, over a certain period of time. This scientific approach helps volcanologists predict eruptions better. Geophysicists pay close attention to the effect of gravity and magnetism on volcanoes.

Geophysicists also try to understand the physical properties of Earth. In addition to studying seismic activity, geophysicists study the heat flow both in the Earth's surface and beneath it, and how the Earth's magnetism affects volcanic eruption.

Experimental Volcanologists

Experimental volcanologists reproduce volcanoes in laboratories, but on a smaller scale, of course. They re-create the moving magma, gas emissions, how far **pyroclasts** have been flung into the atmosphere, and the collapse of a **caldera**. These experiments help scientists understand what happens, and how it happens, in a particular environment, with substances they select. The information they gather is then applied to real-world volcanoes, to help figure out what would happen with a real volcano under a given set of circumstances.

Computer Analysts

Computer analysts sometimes go to the site of volcanoes, but they do most of their work in offices and laboratories. They design systems (computer software and hardware) and models of equations (mathematical formulas that describe a volcano) to learn what happens before, during, and after a volcanic eruption. Computer analysts who work in volcanology are usually physicists or engineers, or sometimes geologists. They study the rise of magma to the Earth's crust, the formation of a **magma chamber**, and the

Other Subdisciplines in Volcanology

Seismologists—A seismologist studies earthquakes, and uses geology and physics to understand their strength and effects.

Stratigraphers—A stratigrapher studies the different layers of the Earth's crust, which give information about volcanic activity in a region.

Hydrologists—A hydrologist studies the effect of water on volcanic activity.

Mineralogists—A mineralogist studies the mineral content of volcanic rock.

physical and chemical processes that go on inside the chamber, including the fracturing of rocks, and the way magma breaks up. Computer analysts can gather and decipher huge amounts of information about volcanoes. Because there is so much complex information about volcanoes, it could not be studied without computers.

Geodesists

Geodesists study the size and shape of the Earth, its gravity, tides, polar motion, and rotation. They apply their knowledge to volcano studies by looking at changes that volcanic eruptions have caused in the shape of the Earth.

Predicting the Eruption of Mount St. Helens

USGS volcanologists studied Mount St. Helens in the 1970s, and knew it had been the most active and explosive volcano in the entire Cascade Range of mountains for thousands of years. Based on their studies, they were able to predict that the next eruption would occur before the year 2000. Their predictions proved correct: Mount St. Helens erupted on May 18, 1980, for the first time in 123 years (Figure 2.3). For days and even months before the eruption, volcanologists provided information to local authorities, helping plan evacuations and providing officials with hazard maps.

The 1980 eruption of Mount St. Helens was one of the largest volcanic explosions in North American history. The northern face of the mountain fell, followed by a sideways blast of stone, ash, and poisonous gas that carried debris 17 miles (27 km), flattening the surrounding forest. Despite the warnings, 57 people died, along with thousands of elk, deer, bears, and coyotes. Some 230 square miles (600 km^2)

Figure 2.3 The eruption of Mount St. Helens in 1980 gave volcanologists a unique opportunity to test the latest equipment in their field.

of forest were destroyed. The volcanic plume rose 80,000 feet (24,400 meters) into the air, turned day into night, and covered a large area of the northwestern United States with volcanic ash. The summit of Mount St. Helens was replaced by a horseshoe-shaped crater 2,460 feet (750 meters) deep. A number of smaller eruptions, beginning on May 25 and continuing for years, into 1986, resulted in lava flows that built up a dome in the crater. The volcano and surrounding area are now part of Mount St. Helens National Volcanic Monument. They have provided biologists with a unique opportunity to observe ecological succession and the reestablishment of natural habitats.

Chapter 3

History of Volcanology

ANCIENT TIMES

Primitive peoples thought volcanoes were omens—signs of things to come. They performed rituals of human sacrifice to ward off the evil they believed caused eruptions. They carved drawings on the walls of their homes and in caves, depicting explosions from cone shapes with wide arcs stretching from their centers, with flying rocks all around.

Back in the fifth century B.C., Greek philosophers wondered what caused volcanic eruptions. They thought powerful winds were blowing inside the Earth, forcing magma to blow out through openings on the surface. Rock, they reasoned, was heated up and melted by friction, as it was forced up through narrow passages in the Earth. This idea was adopted by Greek philosopher and scientist Aristotle (384–322 B.C.), and initially the Romans, too. Eventually, however, Roman philosopher Seneca decided on another explanation. In the first century A.D., he said he ". . . saw volcanoes as giant furnaces where combustion of fossil fuels, such as coal, bitumen and sulphur [took place]. . . ." He believed these substances were burning, forcing volcanic eruptions. Seneca's interpretation of volcanic events became universal, and it remained the accepted understanding throughout the Middle Ages.

MOUNT VESUVIUS AND THE FIRST VOLCANOLOGIST

After 1,000 years of rest, Mount Vesuvius in Italy began to rumble and emit gases before it finally erupted in A.D. 79. The ground swelled around the mountain, and the poisonous gases that drifted up to the Earth's surface off and on for a few years before the eruption killed hundreds of sheep, according to scientists of the time.

On August 24, A.D. 79, the mountain finally exploded. The eruption and the events leading up to it were described in a letter written by a Roman named Pliny the Younger to the Roman leader Tacitus. Pliny the Younger was the nephew of a famous Roman naturalist, Pliny the Elder. In his letter to Tacitus, Pliny the Younger described in detail what he had seen and heard:

> . . . a fearful black cloud was rent by forked and quivering bursts of flame, and parted to reveal great tongues of fire, like flashes of lightening magnified in size. . . . Soon after-

wards the cloud sank down to earth and covered the sea. . . .
I looked 'round: a dense black cloud was coming up behind
us, spreading over the Earth like a flood. "Let us leave the
road while we can still see," I said, "or we shall be knocked
down and trampled underfoot in the dark by the crowd
behind." We had scarcely sat down when darkness fell, not
the dark of a moonless or cloudy night, but as if a lamp had
been put out in a closed room. You could hear the shrieks of
women, the wailing of infants, and the shouting of men. . . .

The notes of Pliny the Younger go on to describe the earthquakes before the eruption of Vesuvius, the **eruption
column**, air fall, the effects of the eruption on people,
pyroclastic flows (Figure 3.1), and even a **tsunami**.
Because of these observations, most volcanologists think of
Pliny the Younger as the first volcanologist.

**Figure 3.1 The pyroclastic flows from the eruption of Mount Vesu-
vius in A.D. 79 preserved the bodies of some of the victims almost
perfectly intact.**

The second part of Vesuvius's blast involved pyroclastic flows that surged toward Herculaneum, destroying this city as well, by burying it under 66 feet (20 meters) of **pumice** and ash, and destroying whatever was left of Pompeii. These cities never recovered from the eruption. They disappeared completely until recent technology enabled scientists to excavate the area.

Vesuvius is a compound volcano. This means that it is actually two volcanoes in one. Monte Somma is an older, collapsed **stratovolcano** with a caldera about 3 miles (5 km) in diameter. Inside the caldera of Monte Somma is the cone of Mount Vesuvius.

The A.D. 79 eruption of Vesuvius was followed by continuous strombolian eruptions, building on the Vesuvius cone. The volcano erupted again in A.D. 472, devastating the areas around it once more. Another eruption occurred in 1139, and again in 1631. These disasters killed 4,000 people and destroyed the area with pyroclastic flows and ash.

The many eruptions and tragedies of Vesuvius led the king of Naples to establish the royal Vesuvian Observatory in 1841, the first volcano observatory in the world. Vesuvius has remained under constant surveillance ever since.

KRAKATAU AND MODERN COMMUNICATIONS

The 1883 eruption of Krakatau, in Indonesia, made the loudest sound ever heard. It could be heard as far away as Australia, 3,000 miles (4,800 km) away, and spurred violent seaquake-created tsunamis that rose 100 feet (31 meters) high. The fallout from the eruption and the tidal wave together killed 36,000 people. The telegraph made it possible for people to communicate the consequences of

Vesuvius Eruptions

On the days before Vesuvius erupted, the city of Pompeii was like any bustling city of its time. People were out and about on the streets and in shops, greeting friends. Merchants dragged teetering carts filled with goods, pulled by donkeys. None of them knew that within a few days, their city would be buried in ash and pumice, causing houses and roofs of buildings to collapse, killing thousands of people. Pompeii was buried under 16 feet (5 meters) of ash and pumice by the so-called Plinian eruption of Vesuvius, which lasted for 19 hours in the year A.D. 79.

the eruption to people far away, and therefore provide a broader basis for later study.

HOPKINS AND PLATE TECTONICS

After a few centuries of scientific study, including the physics and chemical nature of volcanoes, there was a shift in volcanology, due to the work of William Hopkins. Hopkins was a British, Cambridge-educated, mid-18th-century geophysicist. He looked at volcanoes as a study in thermodynamics, the laws of energy and heat. He proposed that melting rock was caused when rock plates were melted through a process of **convection**, in which they are always being rotated, over millions of years, from upper layers, where it is cooler, down to lower layers, where they melt again. When the lower melted rock—the magma—comes near the surface, it occasionally escapes through cracks, and causes volcanoes to erupt. These ideas formed the basis of plate tectonics, and is the foundation of volcanology as we know it today.

PELEE

The eruption of Mount Pelee was significant to the development of volcanology, as two scientists recorded events during and after its eruption.

Pelee, on the island of Martinique in the West Indies, erupted in 1902, killing 30,000 people. Most of the human toll took place in and around the city of Saint-Pierre. The need to evacuate wasn't clear to the people of the city. There had been Earth tremors, but they were anticipating an earthquake, or maybe some streams of lava flowing into the river valleys of Pelee. They did not expect the black cloud that soon erupted, covering their city and killing thousands.

Frank Perret, a former assistant to the inventor Thomas Edison, was an American volcanologist who went to the island to study the pyroclasts. He established an observatory at Pelee, but he also worked in a hut, even as the mud flows surrounded him. He was the first person to report that pyroclastic flows make no sound at all when they move at hurricane speeds.

Thomas Jaggar, a professor of geology from MIT (Massachusetts Institute of Technology), also visited the scene of

Mount Pelee

Mount Pelee is an island arc stratovolcano, which forms lava domes at its summit. The lava domes are destroyed by explosions from within the volcano, producing gas, ash, and lava fragments that flow along the volcano's sides. The 1902 eruption of Pelee gave the field of volcanology a new term—*Pelean eruptions*—which are the type that occur when domes are destroyed by pyroclastic flows.

Mount Pelee in 1902. He was so moved by the destruction and loss of life that he decided to devote his life to understanding volcanoes, so that "no more shall the cities be destroyed."

Throughout the 18th and 19th centuries, there were many advances in the understanding of physics and chemistry, which played an increasingly important role in the study of volcanoes. Geologists studied the relationships between intense pressure, heat, and **radioactivity** in the Earth's crust and the effects on volcanic activity.

As technology improved, volcanologists were able to better predict when an eruption might occur. The invention of the seismometers, seismographs, and, eventually, electronic sensing equipment, such as **tiltmeters**, meant that volcanology was at a turning point.

MOUNT ST. HELENS

In more recent history, the eruption of Mount St. Helens and the years leading up to it provided volcanologists with a living laboratory, where they could try out their new electronic sensing gear and computer programs.

On May 18, 1980, at 8:32 A.M., Mount St. Helens erupted, sending a hot avalanche of ash, dust, magma, and mud at supersonic speed across 232 square miles (601 km^2). The vertical column reached 10 miles up into the sky. Pyroclastic flows followed, covering the valleys nearby. After the eruption, **lahars** further damaged the surrounding forest and nearby river valleys.

Robert Tilling, former chief of the Office of Geochemistry and Geophysics at USGS said, "Mt. St. Helens launched the decade of the eighties for volcanology." Tilling explains that geologists have only recently been able to predict eruptions within a time frame of 10 to 20 years. Newly developed electronic instruments and computer software had enabled scientists to monitor and analyze data from stratovolcanoes, but they had no active volcanoes to test them on, until Mount St. Helens. In the months before and after the Mount St. Helens eruption, they got their opportunity. Only 57 people lost their lives in the eruption of Mount St. Helens. Many more might have died if not for the instruments, investigations, and warnings from the volcanologists.

There are about 600 active volcanoes in the world today. Most of them, "we know very little about," says Tilling. "More volcanoes exist than there are scientists to study them." Today, observatories around the world collaborate with volcanologists of other, less developed nations, to help predict disasters. They have helped make people safer from the hazards of eruptions, but they cannot predict eruptions with perfect accuracy.

ANOTHER WAY OF LOOKING AT IT

The work of Bernard Chouet, a volcano seismologist with the U.S. Geological Survey's Volcano Hazards Team in Menlo Park, California, has received a lot of attention, because Chouet has been able to successfully predict eruptions with accuracy, within days, and sometimes within hours, at three different sites. He does this by observing something he calls "long period seismic events." His technique was tested in December 2000, when he was able to forecast the largest eruption of Mexico's Popocatepetl in 1,000 years, allowing the orderly evacuation of 30,000 people.

Chouet's background in the theories of physics led him to his prediction technique. He had studied the work of a

Japanese seismologist named Takeshi Minakami. Minakami was one of the early scientists interested in making seismic measurements of volcanoes. In describing Minakami's work, Chouet says, ". . . he was mostly interested in classifying seismic events (quakes, rumbles, and movements beneath the Earth's surface), and establishing some kind of order in the richness of his observations. He ended up classifying seismic events based on their character as A-type or B-type events."

An A-type event is the sound of rock breaking as the volcano readjusts itself to the magma movement. A B-type event includes the rock breaking, but what Chouet found buried within the B-type event was something called the long period event. "The long period event reflects the change in flow pattern of the fluid (magma) that is being pushed through the cracks," Chouet says.

Chouet began by trying to interpret long period events at Mount St. Helens in 1981 (after the main eruption). He put a seismometer inside the crater, next to the **lava dome** that was building up inside of it. With a fellow scientist, Chouet noted several A- and B-type events, some with peculiar "signatures." A seismic signature is a pattern that appears on a seismograph that is unique to a particular movement of magma, or rock.

Chouet says that the long period events are important because they reflect a process that involves fluid. Scientists want to know where the fluid (magma) is, what it's doing, and how fast it is building up pressure beneath the volcano. "This is essentially a pressure cooker situation [volcanoes]. The evidence of this . . . comes through the long period events," says Chouet. Long period events have shown, by their signature "squiggles" on the seismometer, whether magma is building up toward an eruption, or going down again.

Chouet has predicted eruptions with his method, with greater accuracy than with any other approach. In December 1989, one of his colleagues invited him to look at some activity volcanologists had picked up from the Redoubt volcano in Alaska (Figure 3.2). They had noticed a pattern of events, which Chouet checked and immediately recognized as long period events. He told the scientists they had an eruption on their hands. The team of scientists was surprised, and told Chouet he should call Tom Miller, the lead scientist in charge of this Alaskan volcano. Miller answered the phone, saying, "I can't speak now because Redoubt is erupting!"

Figure 3.2 Bernard Chouet's unique method of predicting volcanic eruptions helped scientists prepare for the eruption of Mount Redoubt in 1989 and 1990.

Chouet was able to predict another eruption of Redoubt a few weeks later, on January 2, 1990. He noticed the same pattern of long period events accelerating. He told Tom Miller, "I think we're in a similar situation to where we were on December 14, and we're going to have a major eruption on our hands within 24 hours." Tom Miller notified the workers at an oil refinery near the base of the mountain, and told them to evacuate, due to a coming eruption of Redoubt. The mountain exploded just two hours after the evacuation.

Chouet has also predicted eruptions at Mount Galeros, Mexico, Mount Popocatepetl (called Popo, for short) and Mount Voin in Mexico, and Mount Pinatubo in the Philippines, in 1991. In Popocatepetl's 1994 eruption, 50,000 people were evacuated before the mountain blew up. Chouet remains modest, however. He doesn't claim that he has found the ability to predict *every* volcanic eruption. He says that each volcano is different, and should be studied according to its unique history and chemical makeup.

TYPES OF ERUPTIONS

Volcanologists have identified five types of eruptions: Hawaiian, Pelean, Strombolian, Vulcanian, and Plinian.

Hawaiian Eruptions

These eruptions are named for their frequent occurrence in the Hawaiian Island volcanoes. Lava pours out of a crater in a stream, forming lakes.

Pelean Eruptions

Named after Mount Pelee's 1902 eruption, these cause the formation of avalanches, made up of thick magma and part of the volcano's dome (the top of the volcano).

Strombolian Eruptions

These eruptions, the kind that occurred at Mount Stromboli and Mount Etna, are named for the Stromboli volcano of the Aeolian Island. They produce small lava "bombs," rocks, and ash, with the lava pouring out in streams.

Vulcanian Eruptions

These eruptions are produced by an explosion of magma gas. They have repeated explosions with a discharge of large bombs, along with clouds of gas and cinders that rise to around 12 miles (19 km) in the atmosphere.

Plinian Eruptions

Plinian eruptions, like the kind that affected Mount St. Helens and Vesuvius, are the most violent. They produce large volumes of magma, rich in gas. During a Plinian eruption, magma flow can vary between 22 million and 22 billion pounds of volcanic material per second!

MOUNT PINATUBO

In 1991, after nearly 400 years of inactivity, Mount Pinatubo produced one of the largest eruptions of the 20th century (Figure 3.3). A fissure along the northwestern slope of the volcano produced gas and ash. A seismic station northeast of the volcano registered 223 earthquakes. This indicated to the volcanologists that the volcano was fracturing. There were some explosive events on June 12, and they ended on June 15. That morning, a Plinian explosion occurred, producing a column of gas, ash, and pumice that rose 25 miles (40 km) high in the sky. The column spread out, producing an umbrella that spead over a few hundred miles. It blackened the sky and rained pumice and ash over

Figure 3.3 Mount Pinatubo erupted in 1991 after 400 years of being inactive.

more than 115,000 square miles (297,849 km^2). Because of the combined efforts of Filipino volcanologists and USGS volcanologists, who had studied the volcano for months before the eruption, the risk was recognized in time, so fewer people died. However, 883 people were killed, and the economy suffered from the enormous damage. The Clark Air Base, one of the largest in the world, was located nearby. It was destroyed in the volcano's aftermath, and had to be abandoned.

Chapter 4

Where Volcanologists Work

VOLCANOLOGISTS WORK OUTDOORS A LOT. They often set up camp near the volcano, and sleep in tents. Some volcanologists have actually slept inside the crater of a smoldering volcano. When they are at the site, volcanologists work in all kinds of weather conditions—whether cold or hot, raining or snowing. They bring the clothing and supplies necessary for their comfort and survival.

Most of the active volcanoes in the world are along the Pacific Rim, and volcanologists travel to them when they can. Many active volcanoes along the Pacific Rim are in developing nations that don't have their own trained volcanologists for hazard prevention studies. USGS volcanologists and other volcanologists around the world go to these countries to help forecast eruptions and save lives.

They often travel by helicopter to remote areas, or use four-wheel-drive vehicles to get to hard-to-reach sites. Sometimes, they have to make the final trek on foot, because getting through dense underbrush with a vehicle proves impossible. Even on foot, however, the going can be tough, so scientists have to carry machetes (large knives) to cut a path up the mountain to set up their instruments.

GOVERNMENT AND GEOLOGICAL INSTITUTES

Many geologists and volcanologists combine research and field work with jobs at government agencies, like the United States Geological Survey (USGS), the American Geological Institute, the National Aeronautics and Space Administration (NASA), or the National Science Foundation.

Supervolcanoes

Supervolcanoes occur when magma rises from the mantle to create a boiling reservoir in the Earth's crust. The magma chamber increases to an enormous size, building up pressure until it finally erupts.

The last supervolcano eruption, according to scientists, was Toba, 74,000 years ago, in Sumatra, Indonesia. Climatologists say that Toba blasted so much ash and sulfur dioxide into the air that it blocked the sun and caused the air to grow so cold that life on Earth was affected greatly. Some scientists believe that Toba's eruption may have reduced the Earth's population to just a few thousand people.

TEACHING

Volcanologists may teach at colleges or universities and create community programs for educating the public about volcanoes. Volcanologists generally divide their time between office work (or teaching) and laboratory and fieldwork.

OBSERVATORIES

Volcanologists work in observatories, located near active volcanos. Observatories are research centers designed for scientific study of some aspect of nature. Volcanologists work in observatories year-round, monitoring earthquake activity, ground deformation, gas chemistry, and other conditions before, during, and after eruptions.

Volcano observatories are buildings with offices. They are equipped with remote-sensing devices, computers, and a wide range of devices used to monitor volcanoes. When volcanologists work in observatories, they assess potential hazards by identifying and studying past volcanic events. They study the products of these eruptions and areas that would be affected by similar events in the future.

Observatories are a safety net for communities living near volcanoes. They provide warnings during volcanic crises, and they monitor restless volcanoes. They interpret the results of the monitoring and make hazard assessments, create response plans, and design public education programs. Observatory equipment can also be mobilized, so that volcanologists can monitor volcanoes in remote areas, while working on hazard teams.

As part of their jobs in an observatory, volcanologists study the impact of a volcano on the environment, including the effects of volcanic gases on the atmosphere, increased sediment in streams and on the life-forms that depend on them. They do research by reading books and studying lab results.

USGS OBSERVATORIES

There are several observatories in the United States. Alaska Volcano Observatory (AVO) has two locations in Anchorage and Fairbanks. AVO has about 22 full-time scientists, technicians, and administrators.

USGS Cascades Volcano Observatory (CVO) is located in Vancouver, Washington, and is headquarters for the USGS Volcano Disaster Assistance Program. The CVO is the watchdog for Mount St. Helens, which continues to be active since its huge explosion in 1981. It also watches other volcanoes in the Cascades Mountain Range.

USGS Hawaiian Volcano Observatory (HVO) is located in Hawaii Volcanoes National Park, on the Big Island of Hawaii. HVO watches Mount Kilauea and Mauna Loa, two active shield volcanoes located on the Big Island, and the other "hot spot" volcanoes of Hawaii. Mauna Loa is the largest active volcano on Earth.

Long Valley Caldera Observatory has its headquarters at the USGS office in Menlo Park, California. The Long Valley caldera is a 10-mile-wide (16-km-wide) depression in the land along the eastern side of the Sierra Nevada Mountain Range in east-central California (Figure 4.1). This crater was caused by the explosion and collapse of the volcano's cone 760,000 years ago. Long Valley caldera is monitored by instruments placed in and around the site. The data collected go to Menlo Park, where scientists analyze the information.

Figure 4.1 **The Long Valley caldera in California is home to one of the most important observatories in the United States.**

Yellowstone Volcano Observatory (YPO) is located in Yellowstone Park, Montana. It monitors "the largest volcanic system in North America." Within Yellowstone Park, there is a giant caldera, covering an area of 47 by 28 miles (76 by 45 km). The caldera sits over a hot spot, where a supervolcano erupted 600,000 years ago. The YPO monitors the area with seismometers and with **Global Positioning System (GPS)** networks. Recently, there has been an increase in earthquakes in the area, and volcanologists have measured a rise in the land surface over the caldera.

Chapter 5

Tools of the Volcanologist

VOLCANOLOGISTS STILL USE SOME techniques from the old days of volcanology, such as surveying instruments, compasses, and hammers. But today, tools of the volcano trade also include highly sensitive technology. Determining the history of a volcano is always the first step, but after that, volcanologists use seismographs and other types of devices that sense movement and gas.

MEASURING MOVEMENT

A seismograph can register the **magnitude**, **escalation**, and **epicenter** of an earthquake (Figure 5.1). Volcanic earthquakes are caused by magma moving beneath the volcano. To a volcanologist, the more seismographs on hand for picking up movements, the better. Having more information helps volcanologists predict eruptions with greater accuracy.

Volcanologists strategically place the instruments on the volcano, to create a seismic network. The network then sends seismic information to computers by radio 24 hours a day. The computers analyze all the data they receive.

As the activity near a volcano increases, ground swelling and cracks often appear near its edge. This is called ground deformation. It is measured with electronic distance meters (EDM), global positioning systems (GPS), strainmeters, and tiltmeters.

Figure 5.1 A seismograph (the record of volcanic activity printed out from the readings of a seismometer) shows the changes in the Earth that lead up to a volcanic eruption.

Strainmeters and tiltmeters have highly sensitive sensors that note subtle changes in the shape of the ground surface. EDMs use lasers to measure changes in distances between benchmarks (fixed points). Global positioning systems communicate with satellites orbiting the Earth, to determine and track the locations of ground deformations.

MEASURING GAS

As magma rises to the surface, dangerous gases begin to bubble out into the atmosphere. Volcanologists place spectrometers near the base of the volcano to check for sulfur dioxide and carbon dioxide. An increased presence of these gases indicates a possible eruption.

The most widely used instrument for measuring volcanic gas emissions was invented in the 1950s. It's called the "Japanese box," after the Japanese researchers who created it. It consists of beakers of potassium hydroxide, which are placed inside a crate. When volcanic gases interact with the potassium hydroxide, the gases change its composition. This change increases before an eruption.

The Japanese box must be checked in person, which is pretty risky for volcanologists, since it's usually being checked because an eruption is expected at any moment. Dr. Stanley Williams (Figure 5.2) is designing an electronic Japanese box to make it safer to get readings just before an eruption. It will send information to a remote observatory. Williams's instrument uses tiny electrochemical sensors that create currents proportional to the amounts of various volcanic gases in the air.

Another device under development for remote gas monitoring was originally used to monitor toxic gases from factory smokestacks. Called "**correlation spectrometers**,"

Figure 5.2 Dr. Stanley Williams, a prominent American volcanologist, has designed a new type of Japanese box to take readings of volcanic activity just before an eruption.

these devices use an infrared telescope to monitor emissions of toxic gases.

Gas-sensing devices can be mounted in aircraft to determine how much gas is being emitted each day. During an eruption, sulfur dioxide can be measured from space with instruments aboard satellites.

Chapter 6

Safety and Volcanology

IT'S HARD TO IMAGINE a job more dangerous than the volcanologist's. The gases from eruptions are poisonous, even deadly. Explosions and eruptions can occur without warning. The ability of scientists to predict eruptions has come a long way, but even with great advances, there is no way yet to keep volcanologists entirely safe.

Since most volcanoes can erupt at almost any time, volcanologists must always have a plan in mind for escape. If there is a violent eruption, they need to know what to do to survive. Because of these factors, and the large number of unfortunate accidents involving volcanologists, the International Association of Volcanology and Chemistry of the Earth's Interior (IAVCEI) created a subcommittee in March 1993 "to consider procedures that could prevent, or at least reduce, the incidence of such disasters at active volcanoes." The committee made the following suggestions, to increase safety for volcanologists:

1. Scientists doing research on a volcano should have a comprehensive safety plan to minimize hazards and save lives in the event of an unexpected eruption.
2. During planning of the project, the volcanologists should contact local authorities responsible for civil defense, disaster mitigation, and rescue, and the procedures to be taken in case of an emergency should be discussed.
3. The daily work schedule of the field party should be left with local authorities or colleagues who remain outside the hazardous area.

Danger on the Job

June 1991: Three volcanologists and 40 other people were killed when they were engulfed by a pyroclastic flow from the collapse of a lava dome at Unzen volcano, Japan.

January 1993: Six volcanologists and three other people were killed by an explosive eruption at Galeras volcano, Colombia.

March 1993: Two volcanologists were killed by a **phreatic explosion** at the crater of Guagua Pichincha volcano, Ecuador.

4. Volcanologists should contact local researchers, especially where a volcano observatory is in operation, to reduce confusion in case of an accident.

5. Volcano scientists should avoid working alone, and the size and composition of the field party should be optimal for the specific fieldwork. A large group would require a different action plan from what a smaller group would need. Visits to hazardous areas by very large groups, such as field excursions connected to scientific meetings, should be avoided.

6. Do not include inexperienced people such as tourists, reporters, TV crews, and others for travel with scientists into hazardous areas.

7. Volcanologists should have first-aid training and arrange insurance coverage, and should take weather conditions into consideration, while always using common sense.

SAFETY EQUIPMENT

When a volcano blows, volcanologists need the right equipment to wear and to get help (Figure 6.1). The IAVCEI recommends the following safety equipment for volcanologists:

- Handheld, two-way radios
- Protective helmets (hard hats) with chin straps
- Full-face and half-face gas masks (respirators) should always be carried, especially when working in thick fumes or in areas of high gas concentration. The correct type of absorbers and filters, with an ample supply of spares, should be used.

Figure 6.1 These volcanologists are decked out in the heavy-duty clothing needed to protect the body against the heat and gases at a volcano.

- Clothing should be suitable for harsh weather conditions and for protection from falling ash and heat. Brightly colored clothing will increase visibility of field party members and help during possible rescue operations.
- Heavy-duty boots with good ankle support
- Gloves thick enough to provide protection from cuts, abrasions, and burns are essential when working on fresh lava.
- Adequate water and food supplies
- Topographic maps, compass, altimeter, knife, whistle, and a signal mirror may be useful.
- Identification tags or the equivalent, with blood type and the name and address of person to contact, will greatly help in case of serious accidents.

- Goggles or other suitable eyewear may be useful for protecting eyes from blowing ash and corrosive fumes.

USGS HAZARD TEAMS AND VOLCANO DISASTER ASSISTANCE PROGRAM (VDAP)

Volcanologists play a critical role in saving lives around the world. They use mobile observatories for forecasting eruptions and consult with local leaders about hazard prevention and evacuation programs. When the volcano Nevado del Ruiz in Colombia erupted in 1985, its massive lahars killed more than 23,000 people. The USGS and U.S. Office of Foreign Disaster Assistance developed the Volcano Disaster Assistance Program (VDAP) to prevent future tragedies like this one. The program's goal is to respond to volcanic crises around the world, and to reduce fatalities and economic losses.

The equipment used in observatories around the United States and Europe can be moved to other sites. This portability allows volcanologists to give the world their expertise and technological advances. VDAP and its Hazard Teams have saved lives and property.

When invited to do so, VDAP volcanologists work with an at-risk community and its local scientists to help determine "the nature and possible consequences of volcanic unrest."

VDAP's quick response to the 1991 Pinatubo eruption in the Philippines, along with Filipino scientists, "resulted in rapid establishment of a monitoring network and completion of a hazard assessment, accurate eruption forecasts, and effective communication with Philippine Government officials and US military leaders. Evacuations and other civil-defense actions saved thousands of lives and hundreds of millions of dollars of US military aircraft and hardware."

VDAP educates communities around the world, to help people better understand volcanic hazards: ash falls, hot-ash flows (pyroclastic flows), lahars, landslides, tsunamis, lava flows, and volcanic gases.

VDAP describes what its members do, as disaster response scientists:

1. We collaborate with local scientists
2. We deploy a mobile volcano observatory rapidly
3. We assist in unraveling a volcano's history to identify potential types and sizes of future eruptions and to produce volcano-hazard maps
4. We work with other scientists to develop and install "robust" volcano monitoring equipment
5. We apply lessons learned through our international responses to better monitor volcanoes and reduce volcanic risk in the United States.

WORLD ORGANIZATION OF VOLCANO OBSERVATORIES (WOVO)

WOVO is an organization of and for volcano observatories around the world. It includes the Americas, Europe, Africa, Asia, and Antarctica. The observatories work together to monitor volcanoes and are responsible for warning authorities and the public about hazardous volcanic unrest. The key aims of WOVO are:

- To stimulate communication and cooperation between observatories and institutions directly involved in volcano monitoring,

- To develop and maintain volcano monitoring reference materials, including a directory of member observatories, their monitoring networks and staff,
- Upon request, to help a member observatory to find temporary scientific reinforcement,
- To refer governments, international organizations, and others seeking assistance in volcano monitoring to the appropriate member observatories.

Chapter 7

Profiles: Why Did You Become a Volcanologist?

CHRISTINA HELIKER DIDN'T PLAN to become a volcanologist (Figure 7.1). She was working at an office for the USGS in Tacoma, Washington, which was studying glaciers, when Mount St. Helens erupted. She said, "I had absolutely no idea that I ever wanted to be a volcanologist, but the eruption of Mt. St. Helens changed the course of a lot of people's lives, and I was one of them. I went down to St. Helens the weekend after the big eruption and volunteered to help." She was especially moved by the sight of the devastation of Mount St. Helens eruption, and watching "the lava dome grow in the crater over

Figure 7.1 Volcanologist Christina Heliker takes a lava sample at a volcano site.

the next couple years . . . I think any earth scientist who had the opportunity to be there was instantly converted to volcanology."

Volcanologist Thor Thordarson grew up in Iceland, a place that is full of volcanoes. He saw his first volcanic eruption at the age of 6, and checked out some glowing lava on an excursion when he was 12. He explained, "These

Profile in Volcanology

Thomas Jaggar was born in Pennsylvania in 1871, and was fascinated with the natural world as a child. He received a Ph.D. in geology from Harvard University, but his career as a geologist was changed forever when he heard the news of the eruption of Pelee and others killing thousands of people. He wanted to look into the situation for himself. With the help of the U.S. Navy and the National Geographic Society, Jaggar went to Martinique to observe the Mount Pelee volcano, which had killed 28,000 people.

He describes the scene in his notes: "It was hard to distinguish where the streets had been. Everything was buried under fallen walls of cobblestone and pink plaster and tiles, including 20,000 bodies. . . . As I look back on the Martinique experience I know what a crucial point in my life it was. . . . I realized that the killing of thousands of persons by subterranean machinery totally unknown to geologists . . . was worthy of a life work."

In 1909, in response to the tragedy he'd witnessed, he established the first volcano observatory in the world, located in Hawaii, studying Mount Kilauea. In doing so, he furthered the science of predicting volcanoes, and prevented loss of life.

experiences must have had some effect, because the images . . . are still as clear today as they were then." He says the decisive moment for becoming a volcanologist came in August 1983, "when I first laid my eyes on the Laki cone row (a row of cone volcanoes immersed in orange red light of the setting mid-summer sun). . . . At that time I was completely taken by the urge to understand these sublime, and at the same time mysterious or even mythological structures."

For Kathy Kashman, it was the sight of red lava that made her want to become a volcanologist. She got her chance while she was part of a monitoring program at Mount Erebus in Antarctica (Figure 7.2), the world's southernmost volcano: "It was amazing because the volcano is thirteen thousand feet-high, and it was probably 20 or 30 below zero on the top. There is a great big crater in the top of the volcano with very, very steep sides. We would walk up to the top and lie down on our stomachs and look down. If you looked way down you could see the lava lake—red lava with a black crust, and the whole lake circulating and convecting. You could feel the heat from it."

Planetary volcanologist Jeff Byrnes developed his career by simply following his interests. As a college student, he majored in environmental geology, but then he took a class called "Planetary Materials." This led him to studies of lava flows on Mars. In a 2003 interview with the National Science Teachers Association (NSTA), Byrnes said these experiences sparked his interest in volcanology on other planets. "I began my graduate studies in planetary volcanism the following fall," he said. Today, Byrnes does research both on Earth's volcanoes and volcanoes on other planets. He enjoys his work for many reasons. "First and

Figure 7.2 Mount Erebus in Antarctica, seen here in the background, helped convince Kathy Kashman to make volcanology her career.

foremost," he says, "I find volcanoes incredibly interesting and, as a field-oriented volcanologist, I enjoy traveling to see beautiful volcanoes around the world. Additionally, I have the opportunity to experience various cultures of people living in volcanic areas."

Chapter 8

New Advances in Volcanology

A NEW TECHNOLOGY IS MAKING it possible for volcanologists to detect volcanic activity on Earth from space. By using a technique called InSAR (interferometric synthetic aperture radar) scientists can detect even tiny bulges, as small as a few inches, from satellites orbiting Earth (Figure 8.1).

Weather satellite pictures are now being used by volcanologists to monitor 100 dangerous volcanoes along the Pacific Rim in Alaska and Russia. They are looking for excess heat (with heat sensors) that indicates a volcano is likely to erupt. According to one volcanologist, "The method allows scientists to

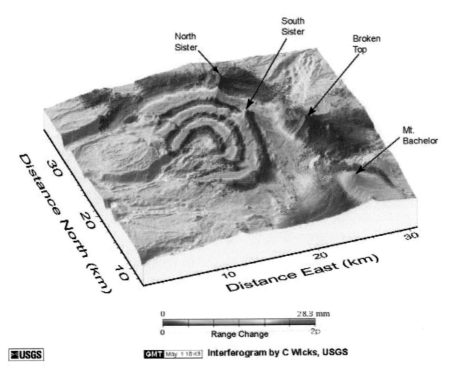

Figure 8.1 This InSAR image of the Cascades area shows mountains called "North Sister," "South Sister," and "Mount Bachelor."

observe volcanoes when it is too expensive to install earth-quake sensors to listen for signs of imminent eruptions."

VOLCANOLOGY IN OUTER SPACE

Volcanology is an expanding field. **Planetary volcanology** and **astrogeology** are new areas of volcano research. Scientists are taking research from volcanoes on Earth and applying this knowledge to active and dormant volcanoes on other planets and moons in our solar system.

Space probes traveling to Mars, Venus, and Jupiter's volcanic moon Io are equipped with remote-sensing devices. They relay information to scientists back on Earth about surface features and underlying ice, water, and magma flows.

The USGS in Flagstaff, Arizona, has been studying volcanism on the moon, as part of the Lunar Pyroclastic Volcanism Project, funded by the NASA Planetary Geology and Geophysics Program. Scientists study the composition of lunar pyroclastic deposits. Lunar pyroclastic deposits are the products of volcanic eruptions on the moon.

Volcanologists have been looking at a lunar deposit called Taurus-Littow, among others, and have found that its dark color is caused by the presence of black beads made of crystallized ilemite. "Pyroclastic materials at this site and others on the moon include volcanic glass beads in a variety of colors," says Lisa Gaddis, principal investigator at the USGS astrogeology program. "[T]hey represent the most basic or primitive lunar materials."

Scientists have obtained this information with remote sensing. "With the global coverage provided by the *Clementine* multispectral data, we now have the opportunity to fully characterize lunar pyroclastic products," according to Gaddis. *Clementine* was a spacecraft launched in 1994. It mapped the 38 million square kilometers (14.7 square miles) of the moon at 11 different wavelengths (in the **electromagnetic spectrum**).

The recent discovery of active volcanoes on Mars (Figure 8.2) has also given rise to a lot of scientific study. In "Flying Above Mount Olympus," from NASA's Web-based publication *Astrobiology Magazine*, a staff writer describes how the European orbiter, *Mars Express*, took high-tech photos of Olympus Mons, the biggest volcano in our solar system. Olympus Mons is 13 miles (21 km) high with a caldera 1.8 miles (2.9 km) deep.

Figure 8.2 Volcanoes like this one have been found on Mars, opening up a whole new area of research for volcanologists.

University of Buffalo volcanologist Dr. Tracy Gregg is quoted in the article, discussing the scientific appeal of studying Martian volcanoes in detail: "If both of these [*Opportunity* and *Spirit*] landers survive with airbag technology, then it blows the doors wide open for future Mars landing sites with far more interesting terrain. A landing site near a volcano might be possible, now that the airbag technology has worked so wonderfully."

"I'd like to see us land ON a volcano," said Gregg. "Right on the flanks. Often the best place to look for evidence of life on any planet is near volcanoes."

Chapter 9

Becoming a Volcanologist

CAREERS IN VOLCANOLOGY require backgrounds in one or more of the natural sciences, including chemistry, physics, and geology, as well as computer science and math. If you might like to become a volcanologist, you should take as many math, chemistry, and physics classes as you can. Depending on their choice of subdiscipline, volcanologists might use math to calculate the density of volcanic deposits, or to chart the flow of volcanic ash over an area. Chemistry and physics are important to volcanology, because they focus on the basic materials of nature, which includes volcanoes. You'll need a college degree if you want a job in volcanology. In fact, almost all volcanologists have at least some graduate education, and most volcanologists have a Ph.D.

According to Edward P. Klimasauskas at the USGS, many universities "provide excellent educational opportunities that could put you on track to becoming a volcanologist. The choice of undergraduate and graduate level study really depends on your individual interest, as each university has strengths and weaknesses."

JUMPSTART YOUR VOLCANOLOGY CAREER

Doing a little research at the library can give you a headstart on your career. You can also contact professors, by sending them an e-mail or writing a letter. The USGS recommends reading about the school you're interested in first so that you will be able to ask informed questions.

Internships

Internships provide real-world experience. Volcanology and geology interns can get a sense of what it would be like to be a professional scientist on a day-to-day basis. Placement offices at colleges can provide internship information. Some programs pay their interns, while others only provide college credit for the work. USGS observatories offer internships to qualified students.

Cascades Volcano Observatory, for example, recently offered the Jack Kleinman Internship for Volcano Research. It provided $500 to $2,000 to seniors in college and graduate students who wanted to conduct research in volcanology, either in the Cascade Range, Aleutian volcanic arc, Hawaii, Yellowstone, or Long Valley caldera. The money was provided to help the interns pay for travel to the field area, living expenses while working, or to purchase supplies.

If you think you might want to become a volcanologist, you should find out all you can about the opportunities that are available for training and education.

Who Becomes a Volcanologist?

Do you enjoy chemistry experiments, or playing with a chemistry set? Have you participated in your school's science fair? Do you enjoy collecting rocks and minerals? These are all things that most geologists and volcanologists enjoyed when they were young.

Often, careers in science begin with exploring the natural world. According to University of North Dakota volcanologist Jessie Yellick, there are three basic traits common to most volcanologists: "First . . . an interest in the outdoors. Backpacking, hiking, and climbing are common hobbies found among this elite and rare group of geologists." He quotes a colleague who says, "Working with volcanoes appeals to the kid in you; the excitement, the danger, the thrill of watching things blow up. Second, a vast interest, and a longing for adventure. . . . Third, some people find that the sense of mystery associated with scientific studies is intriguing as well." He adds that even though the work is fascinating and fun, volcanologists are expected to monitor volcanoes with "keen scientific observation and judgment. You must have the willingness to work hard. . . ."

Volcanologists are good at communicating ideas, putting together different types of information, and forming conclusions. They're good at math, and can multiply, add, and divide quickly and with accuracy. A volcanologist is sharp!

Ash: Small particles of material from eruptions that are smaller than 0.08 inches (2 mm). It isn't dust-like, such as ash from a fireplace. Volcanic ash resembles small marbles, or ground glass.

Ash plume: The tall clouds of thick smoke and ash that hover over the erupted volcano.

Astrogeology: The scientific study of the composition, structure, and physical properties of planets, moons, asteroids, and comets.

Black smoker: A kind of vent found in the ocean floor.

Caldera: Large depression produced by an eruption and collapse of a magma chamber; it is usually round or horseshoe-shaped.

Climatologists: People who study changes in climate and try to predict long-term changes.

Cone: The outer part of a volcano that resembles a cone in shape.

Convection: A method of heat transfer, with hot material rising and cool material falling.

Correlation spectrometers: Infrared telescope devices used for measuring gas emissions.

Crater: The mouth of a volcano.

Deformation levels: The amount of change in the shape of the ground as magma forces its way upward.

Deposits: Rock, clay, sand, and mineral material deposited on the Earth's surface by a volcanic eruption.

Dormant: Inactive, but still alive.

Electromagnetic spectrum: The full range of electromagnetic energy waves. Energy travels through space in waves of different frequencies.

Electromagnetism: Magnetism produced by electric current.

Epicenter: The area just above the source of an earthquake.

Eruption column: The cloud of gas that rises straight above the crater during an eruption.

Escalation: How fast the volcanic eruption is emerging to the surface.

Extinct: A volcano that has died out and will not erupt again.

Fissure: A crack in a rock plate through which volcanic material escapes.

Fumerole: An opening where volcanic gases escape.

Global Positioning System (GPS): A network of satellites and electronic receiving devices that allow the exact location of something to be determined with great accuracy.

Granite: A low-density magmatic rock that makes up continents.

Gravity: The force that pulls objects toward the center of the Earth.

Lahar: An Indonesian word for volcanic mudflow formed when eruptions melt part of a glacier on a volcano; a large flow of volcanic material that moves at high speeds.

Lava: Magma that has reached the Earth's surface.

Lava dome: The rounded area within a caldera, where lava slowly builds up before an eruption.

Lava fountains: Exploding lava shooting high into the air over a volcano as it erupts.

Lithosphere: The rigid outer layer of Earth.

Magma: Molten rock material.

Magma chamber: The part of a volcano where magma accumulates and is stored, a few miles below the Earth's surface.

Magnetometer: Instrument used to observe changes in the magnetic fields beneath a volcano.

Magnitude: The size of an eruption.

Mantle: The layer of Earth beneath the lithosphere.

Molten: Melted.

Phreatic explosion: A type of explosion when magma meets groundwater, which produces a huge amount of steam very quickly, creating a lot of pressure that causes an explosion.

Planetary volcanology: The study of volcanoes on planets and moons in outer space.

Pumice: Foamy volcanic rock.

Pyroclastic flows: Mixtures of volcanic materials and gas traveling at high speeds down the sides of a volcano.

Pyroclasts: Rock material broken into small pieces by volcanic blasts.

Radioactivity: Energy emitted in the form of particles or rays.

Seismic activity: Activity caused by movement of rock, as magma moves beneath the surface of the Earth.

Seismometer: A device for measuring earthquakes.

Spectrometers: Instruments used to monitor volcanic gases.

Stratavolcano: A volcano made up of layers of different materials, from lava and pyroclastic flows.

Subduction zone: A deep trench between tectonic plates where lithosphere is melted.

Sulfur: A common volcanic gas, which gives off a smell similar to that of rotten eggs.

Tectonic plates: Giant slow-moving slabs of rock that make up the Earth's crust.

Tiltmeter: An instrument for measuring ground movement near a volcano.

Trace elements: The smallest chemical quantities of a particular magma sample.

Tsunami: Giant sea wave caused by ocean earthquakes and collapsing volcanoes.

Vent: The passageway through which magma makes its way to the surface.

Viscous: Having a thick fluid consistency.

Volcanologist: A person who studies volcanoes.

Books

Environmental Scientists and Geoscientists: Occupational Outlook Handbook, 2004–2005 Edition. Indianapolis: Jist Publishing, Inc.

McGuire, Bill. *Raging Planet*. London: Barron's Educational Series, 2002.

Rosi, Mauro, Paulo Papale, Luca Lupi, and Marco Stoppato. *Volcanoes*. Buffalo, NY: Firefly Books, 2003.

Sigurdsson, Haraldur. *Melting the Earth: History of Ideas on Volcanic Eruptions*. New York: Oxford University Press, 1999.

Thompson, Dick. *Volcano Cowboys: The Rocky Evolution of a Dangerous Science*, New York: St. Martin's Press, 2000.

Websites

"Flying Above Mt. Olympus." *Astrobiology Magazine*. Available online at *www.astrobio.net/news*.

Information for Future Volcanologists: Questions and Answers on Career Planning. Available online at *http://vulcan.wr.usgs.gov/Outreach/StudyVolcanoes/career_planning.html*.

Lunar Pyroclastic Volcanism Project—Lisa Gaddis. Available online at *http://astrogeology.usgs.gov/Projects/LunarPyroclasticVolcanism*.

Occupational Outlook Quarterly. "You're What? A Volcanologist— Michael Stanton: An Interview With Robert Tilling." Available online at *www.findarticles.com*.

The Power of Volcanoes: Mount Pelee. Available online at *http://library.thinkquest.org/17457/volcanoes/effects.pelee*.

Profiles from Tari Mattox Interviews. *Volcano World Site*. Available online at *http://volcano.und.nodak.edu/vwdocs/interview*.

Satellites Reveal Volcano Dangers—Lee Siegel. Available online at *www.space.com/scienceastronomy/planetearth/volsat1*.

Supervolcanoes. Available online at *www.bbc.co.uk/science/horizon*.

Supervolcano Questions. Available online at *http://volcanoes.usgs.gov/yvo/faqs1.html*.

Tyson, Peter. "Can We Predict Eruptions?" *NOVA Online*. Available online at *www.pbs.org/wbgh/nova/vesuvius*.

U.S. Department of Labor and Statistics, Environmental and Geoscientists. Available online at *www.bls.gov/oco/ocos050.htm*.

U.S. Geological Service. Frequently Asked Questions About Volcanoes and Volcanologists. Available online at *http://lvo.wr.usgs.gov/FAQ.html*.

U.S. Volcano Observatories and the Volcano Disaster Assistance Program (VDAP). Available online at *http://volcanoes.usgs.gov/about/Where/WhereWeWork.html*.

Volcanoes Talking: An Interview with Bernard Chouet. Available online at *www.pbs.org/wgbh/nova/volcano/chouet.html*.

Volcanologists Take to Space—Andrew Bridges. Available online at *www.space.com*.

Volcano SWAT Team NOVA Online. Available online at *www.pbs.org/wgbh/nova/vesuvius/team*.

What Are Some Personality Traits—Jessie Yellick. Available online at *http://volcano.und.nodak.edu*.

Williams, Dr. Stanley. "Career as a Volcanologist." Available online at *http://teacher.scholastic.com/researchtools.articlearchives/volcanoes.career.htm*.

Books

Branley, Franklyn M. *Volcanoes*. New York: Thomas Y. Crowell Junior Books, 1985.

Clarke, Penny. *Volcanoes*. Danbury, CT: Franklin Watts (Grolier Publishing), 1998.

Hayhurst, Chris. *Volcanologists*. New York: Rosen Publishing Group, 2003.

Lampton, Christopher. *Volcano*. Brookfield, CT: The Millbrook Press, 1991.

Lindop, Laurie. *Probing Volcanoes*. Brookfield, CT: Twenty-first Century Books, Millbrook Press, 2003.

Websites

Information for Future Volcanologists: Questions and Answers on Career Planning. Available online at *http://vulcan.wr.usgs.gov/Outreach/StudyVolcanoes/career_planning.html*.

Occupational Outlook Quarterly. "You're What? A Volcanologist— Michael Stanton: An Interview with Robert Tilling." Available online at *www.findarticles.com*.

U.S. Geological Service. Frequently Asked Questions About Volcanoes and Volcanologists. Available online at *http://lvo.wr.usgs.gov/FAQ.html*.

Williams, Dr. Stanley. "Career as a Volcanologist." Available online at *http://teacher.scholastic.com/researchtools.articlearchives/volcanoes.career.htm*.

PICTURE CREDITS

ABOUT THE AUTHOR

MARY FIRESTONE grew up in North Dakota. She lives in St. Paul, Minnesota, with her 11-year-old son, Adam, their pet beagle, Charlie, and cat, Rigley. She has a bachelor's degree in music from the University of Colorado at Boulder, and a master's degree in writing from Hamline University. When she isn't writing articles for magazines and newspapers and books for children, she enjoys gardening and spending time with her son.